THE LIBRARY OF
FUTURE ENERGY

WIND POWER
OF THE FUTURE
NEW WAYS OF TURNING
WIND INTO ENERGY

BETSY DRU TECCO

THE ROSEN PUBLISHING GROUP, INC.
NEW YORK

Published in 2003 by the Rosen Publishing Group, Inc.
29 East 21st Street, New York, NY 10010

First Edition

Library of Congress Cataloging-in-Publication Data

Tecco, Betsy Dru.
Wind power of the future : new ways of turning wind into energy / by
Betsy Dru Tecco.
    p. cm. — (Library of future energy)
Summary: Presents the pros and cons of using wind power to help fight
air pollution and meet the growing demand for electricity.
Includes bibliographical references and index.
ISBN 0-8239-3665-1
1. Renewable energy sources—Juvenile literature. 2. Wind power—
Juvenile literature. [1. Wind power. 2. Power resources.]
I. Title. II. Series.
TJ808.2 .T43 2002
333.9'2—dc21
                                                    2002002511

Manufactured in the United States of America

CONTENTS

INTRODUCTION

Electricity powers our heat and light. It runs our televisions, stereos, and computers. Most of us in the United States have always enjoyed the conveniences of electricity. We expect things to work at the flip of a switch or the push of a button. Rarely do we wonder about the fuels that power electricity.

Today we know that the primary sources of energy that we've depended on for decades—fossil fuels like coal, oil, and natural gas—won't last forever. In the winter of 2001, millions of Californians experienced power shortages that left them with less electricity and bigger electric bills.

We've also learned that when we burn fossil fuels, heat-absorbing gases, like carbon dioxide, are released into the air. Called

This is an aerial view of Los Angeles, California, at night. During the winter of 2001, Los Angeles businesses and residents faced rolling blackouts when utility companies turned electricity off in different areas for limited amounts of time.

greenhouse gases, these fumes trap heat in our lower atmosphere, causing Earth's temperature to rise. Scientists predict global warming will produce drastic results, including coastline flooding from rising sea levels, increased drought and famine, the spread of disease, more severe storms, and even the potential for an ice age. Acid rain, which poisons our air and water, is another environmental danger that comes from using fossil fuels.

With the rise of utility costs, electric power shortages, and growing environmental concerns, what can we do to meet our energy needs for today and for the future?

Countless surveys reveal that Americans want to begin using renewable energy because it is clean and nonpolluting. Energy from sunlight, water, and wind are all considered "renewable" because they constantly regenerate.

In 2000, a national public opinion survey conducted for the Sustainable Energy Coalition found that 51 percent of the 1,105 respondents thought the Department of Energy's research and development budget should focus its funding on renewable energy, energy efficiency, and conservation. 30 percent of respondents favored continued use of fossil fuels or nuclear energy for the bulk of our energy, while 69 percent thought electricity providers should be required to generate a portion of their power from renewable energy sources.

In a Gallup poll taken in the spring of 2001, 91 percent of the 1,005 Americans surveyed favored investments in new energy sources, including the wind and sunlight. An ABC News and *Washington Post* poll indicated that 90 percent of the 1,004 participants wanted the U.S. federal government to develop more ways to use solar and wind power.

Meanwhile, most of our energy comes from coal, oil, and natural gas. Nuclear power, capable of producing dangerous radioactive waste and causing deadly accidents, supplies about 20 percent of the power in the United States. Today, only 10 percent of our power comes from renewable energy sources. And less than 1 percent of that is from wind power! Despite their benefits, renewable energies are usually treated as second-string players.

Only when fossil fuels are scarce, prices skyrocket, and our economy suffers, do people consider using alternative energy. But as public demand and government support increase, the market for renewable energy will expand.

Wind power holds tremendous promise for the future. Read on to learn how wind energy is shaping our lives and the world around us.

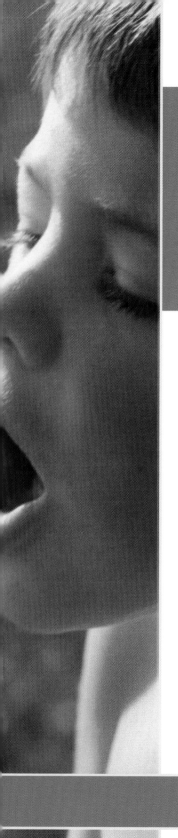

Wind is a wonder of nature that touches every person in every part of the world. Think about how wind can whip at your hair and push your body. If you've flown a kite or sailed a boat, you've felt some of the wind's power. And if you've ever been in a hurricane or a tropical storm, or looked at pictures of damage they've caused, you've seen wind's fury.

WINDMILLS

Wind is moving air. It is created when sunlight warms one part of Earth faster than another, causing differences in air pressure. As air heated by the Sun rises, cooler air sweeps in to take its place. This movement causes wind.

This windmill pumps water to the dry regions of Namibia, a country in southern Africa.

Inventors and engineers have built a variety of state-of-the-art machines called wind turbines, which take the flow of the wind and turn it into energy we can use for practical purposes. Windmills generate mechanical energy for tasks such as pumping water, grinding grain, and chopping wood. In fact, the word "windmill" originated from using these machines to mill, or grind, grain.

More than one million water-pumping windmills are in use around the world. In the rural United States, farmers and ranchers use them mainly to draw water from deep wells for fields and livestock.

Resembling a giant toy pinwheel, the traditional farm windmill has a large rotor with many slanted blades, or sails, extending from a horizontal shaft. Mounted on a tower, the rotor extends up to where the wind is strong and steady. A tail vane (or fantail), like the rudder on a boat, directs the blades into the wind. As the blades revolve in the wind, gears inside the rotor turn, operating a piston pump that draws water up from underground. The water goes into a storage tank or directly onto dry fields.

Heat and Electricity from Wind Power

Wind energy has been used to produce heat by powering wind furnaces. A wind-powered heating system can work in two different ways, although neither method has yet proven economical. It can warm water by driving a paddle or a pump to churn water inside a closed, insulated tank, thus converting mechanical energy into heat. Or, it can produce electric-powered heat by directly driving a heat pump.

The most popular application of wind power, of course, is to produce electricity. From charging batteries to powering entire communities, there are countless uses for wind-generated electricity.

Horizontal-Axis Turbines

The most common wind turbine is called a horizontal-axis turbine. Resembling a water-pumping windmill, it has a fanlike rotor perched horizontally atop a tall tower. With two or three airfoil-shaped blades, the turbine's rotor is similar in design to an airplane propeller. Although usually made of fiberglass, sometimes the rotor blades are made of steel or wood.

When the wind passes over both surfaces of the curved blades, the difference in air pressure between the top and bottom surfaces creates a force called aerodynamic lift. On an airplane, this force

Pictured are wind turbines on a wind farm near Palm Springs, California. Because wind speed is higher and less turbulent at about 100 feet above the ground, wind turbines are usually mounted on tall towers to capture the most energy.

causes the wings to rise, lifting the plane off the ground. On a wind turbine, the same force causes the blades to turn. In addition to lift force, a "drag" force acts to slow the blades. The lift force is much stronger than the drag.

A large blade can capture and produce more electricity than a small blade. A bigger blade turns more slowly than a small blade. In the past, medium-sized turbines, which have blades between 25 and 100 feet long, were more economical and reliable than large turbines, with blades up to 150 feet long. New advances in technology, however, favor the larger turbines for efficiency.

THE NACELLE

The rotor is attached to the nacelle, which sits on the tower and holds all the internal mechanical parts. Some nacelles are large enough for a technician to stand inside while working on the turbine!

The nacelle contains a low-speed shaft. The spinning blades turn the shaft, which goes through a gear transmission box and connects to a high-speed shaft. The gears increase the speed of the shaft so that the generator can make electricity. Although the generator regulates the flow of electrical power, varying wind speeds can make it difficult to keep this kind of turbine running at a constant speed, which is necessary for consistent power.

THE VERTICAL-AXIS TURBINE

The vertical-axis turbine, so called because its blades rotate vertically to the ground rather than horizontally, is another type of wind machine. Resembling a huge eggbeater or a wire whisk with two curved blades, the Darrieus turbine is one example of a vertical-axis turbine. These machines can accept wind from any direction, are easier to maintain, and stand up to high winds. However, because they produce less electricity than horizontal types, vertical-axis turbines are not popular.

VARIABLE-SPEED GENERATORS

Variable-speed generators continue to operate as the wind slows down or speeds up. Superior to constant-speed, or fixed-speed

This photo shows "eggbeater" windmills on a wind plant in Altamont, California.

generators because they can convert more wind energy into electricity, variable-speed generators also eliminate the need for, or reduce the size of, expensive gearboxes.

TURBINE CONTROLS

Some turbines have sensors to measure wind speed and direction. When the data is sent to a computer in the rotor, the rotor triggers a motor that keeps the blades turned in the direction of the wind. Other turbines use a tail vane, or fantail, instead of a computer to stay in the face of the wind. A brake in front of the generator stops the blades from turning if the wind reaches damaging speeds, usually around sixty-five miles per hour. In machines without brakes, the blades can be angled to keep them facing the wind.

WIND FARMS AND POWER PLANTS

Wind turbines are often installed in a group called a wind farm or a wind power plant. The best locations for wind farms have plenty of wind and few obstacles to interfere with airflow. Coastlines, high plains, and mountain passes are ideal. There are several wind plants in northern California's Altamont Pass, the largest developed wind site in the United States. Thousands of large wind turbines line the mountaintops, spinning and whirring like fans perched on tall pedestals. Unlike fans, of course, the turbines don't create wind; they capture it.

Another excellent site for wind plants is offshore in the shallow waters of oceans. The air above the ocean is windy and free of obstructions. Mounted on the ocean bottom, at a depth of sixty feet or less, offshore turbines are becoming more and more common in Europe where there is less land on which to build.

DISTRIBUTING WIND POWER

Wind plants are usually connected to the same electricity transmission and distribution system as are other power plants. They become part of the energy mix. This system is composed of overhead power lines that make up a utility grid. Poles or metal towers called pylons support the lines. The power produced by the utility-scale turbines (those connected to a grid) is collected and delivered over the high-voltage power lines to homes and businesses miles away.

If there are no power lines near the wind plant, new ones must be installed, which is quite costly. Underground lines, which are even more expensive, are normally reserved for short distances.

Vertical-axis turbines turn to generate electricity around the edges of farm fields in Altamont Pass, California.

The performance of a wind plant depends largely on wind speed. The wind at a potential site is evaluated for at least a year, usually with cup anemometers which measure the speed of the wind. The instrument's cups revolve on an upright spindle attached to a speedometer. The greater the wind speed, the faster the cups turn.

When wind speed doubles, the available energy increases eight-fold, so a site with wind speed just a bit higher than another site can provide a lot more wind energy. For example, a site with an annual average wind speed of about 12.6 miles per hour has twice the energy available as a site with an average wind speed of 10 miles per hour.

Utility-Scale Wind Turbines

In general, utility-scale wind turbines require annual average wind speeds of 11 to 13 miles per hour. Nine-mile-per-hour winds are sufficient for turbines that are not connected to a grid, as well as for turbines used solely for mechanical tasks such as water pumping.

Of course, the flow of wind changes with the time of day, the season, or even the type of terrain. Obstacles, like mountains, tall trees, or buildings, affect air movement. Height is a factor, too, since wind speed generally increases (and turbulence decreases) the greater the height from the ground. Many large turbines are about 200 feet tall. Some are even as high as 330 feet tall when measured from the base of the tower up to the tip of a blade pointed toward the sky. That's about as tall as a thirty-three-story building!

Storing Wind Power

It can be tricky to store electricity generated by a large wind turbine. You cannot pour wind into a container to be used later. Engineers have devised a way to get around this limitation using hydroelectric power. On very windy days, when extra energy has been generated, the excess electricity can be used to pump water from a low-lying lake up to a higher lake or reservoir. Then, on calm days, when the demand for electricity exceeds the supply, water from the higher lake can be released down a tunnel to turn the turbines. Because of the expense, this method is rarely used.

The electricity generated from small wind turbines, on the other hand, can be stored in batteries for future use. Homeowners can purchase a small wind turbine to reduce the amount of electricity they buy from the local utility company. The batteries can also provide a backup system for use when the power goes out. (Without batteries, a house that's connected to the utility grid could access power from the electric company when the wind turbine isn't producing enough.) With rotor diameters between three feet and fifty feet, the small turbines generate up to forty kilowatts, which can provide some or all of a household's power. How much power it can supply depends on the wind speed and the monthly power usage for a particular home. It takes fewer than seven years on average to recoup the cost of a small wind turbine.

Villages in remote areas that may not be connected to a utility grid can use groups of small turbines to make power. The electricity generated from small wind turbines is also used for making ice, as well as pumping and purifying water.

The micro wind turbine, small enough to hold in your arms, is the most common kind of wind turbine. Able to produce up to ten kilowatts of electricity, a micro generator can be used to charge batteries on a sailboat, yacht, or in a recreational vehicle (RV). The batteries store the wind-generated electricity to power lights, appliances, and navigational equipment for boats out at sea or RVs on deserted back roads. Micro wind turbines can replace both gas, or diesel-powered generators, which become expensive to operate when fuel prices rise.

COMBINING TECHNOLOGIES

A hybrid system combining wind power with another technology, such as solar electricity (photovoltaic), or diesel electricity can ensure that power is constantly available to people without a grid-connected wind turbine. In a wind/solar hybrid system, for example, the solar component will produce electricity when the wind turbine isn't working on calm but sunny days. When neither the wind nor the Sun can provide enough energy for electricity, batteries and/or an engine-generator powered by fossil fuels can take over the job.

Wind technology continues to evolve and expand. But before we look forward to its future, let's look back at the colorful history that brought wind power to this point.

HISTORY AND DEVELOPMENT OF WIND POWER

For centuries, creative minds have found ways to capture the energy of wind and put it to good use. Egyptians in the fourth century BC used wind pressing against the sails of a boat to transport people across rivers and seas. Sailing vessels sparked long-distance travel as well as commercial trade and the mixing of cultures.

EARLY WINDMILLS

It is not certain when wind was first used for mechanical energy, although there is proof that around the ninth century AD in Persia (now Iran), windmills were used to grind corn for flour and to raise water from streams to irrigate fields. People fastened paddles made of reeds to vertical frames

The sail is an early example of how humans have harnessed the power of the wind.

that rotated in the wind, much like a revolving door. To increase the wind's force, walls were built around the sails to create a wind tunnel.

During the twelfth century, France constructed tide mills to create energy from the ocean tides. The water mills caused arguments over who owned the waterways. Eventually, people in England focused on wind, which no one could own, and built the post mill. A far cry from the Persian windmill, the post mill resembled a small, wooden house set on a revolving platform atop a vertical post. Large paddles covered with cloth were connected to the house by a horizontal windshaft and were positioned to face the wind. When the wind changed direction, the miller used a long pole to turn the entire windmill.

Use of the post mill spread throughout Europe, enabling many people to capture wind energy for a variety of mechanical jobs, from pumping water and grinding grain to sawing lumber and pressing seeds for oil.

Tower mills were developed toward the end of the fourteenth century. Widely used by the Dutch, the tower mill eliminated the need to turn the entire windmill. Instead, only a rotating tower cap, to which the rotor was attached, needed to be turned so that the blades faced the wind. Bigger and sturdier than post mills, the tower mills had room to store the goods they produced.

Windmills became the power behind grinding spices, cocoa, dyes, paint pigments, and much more. It has been estimated that windmills supplied as much as 25 percent of Europe's industrial energy from the 1300s to the 1800s. The balance came from hydropower (water power) as well as from human and animal labor.

During the span of 500 years, windmills helped spark the Industrial Revolution as Europeans enlarged and improved their efficiency. In 1750, a multiblade fantail was invented that automatically kept the rotor pointing into the wind. Other innovations included air brakes; shutters on the blades that adjusted for wind speed; and an aerodynamic blade design similar to modern aircraft wings.

New York City in seventeenth- and eighteenth-century America was home to a cluster of tower mills built by Dutch immigrants. English settlers built smock windmills (their color and shape

Post mills were important to early farming economies in Europe.

resembled a woman's white smock) and post mills along the windy eastern coastline from Cape Cod to the Carolinas. Again, the sturdy machines ground grain and sawed wood.

While windmills gave way to steam engines in Europe during the 1800s, they caught the attention of American homesteaders determined to settle the dry and windswept western frontier. The American farm windmill, which was perfect for pumping water from deep below the earth, became a necessity for people who settled out West. During long periods of drought, the windmills were key to survival. Water-pumping windmills were quite different from tower mills.

COMMERCIAL WINDMILLS

Designed for commercial use in 1854 by Daniel Halladay, these windmills were small, light, multibladed, cheap to build, and easy to maintain. What's more, Halladay built them to withstand the gusty winds of the Great Plains.

For 100 years, reliable farm mills evolved from using blades made from wooden slats to using stamped sheet-metal "sails" that nearly doubled the rotor's efficiency. Power windmills were developed to grind grain and perform many other tasks, such as crushing ore and churning butter. Eventually, water-pumping windmills enabled people to have indoor plumbing for the kitchen and the bathroom. It has been estimated that between 1880 and 1930, six million American windmills were erected on the Great Plains and in the American West.

WIND CHARGERS

Meanwhile, farmers and ranchers also relied on wind chargers, which were designed to charge batteries for radios and other home appliances. Wind chargers, or wind generators, were also used to provide light for remote airplane runways and to power lighthouses. A wind generator developed by the Jacobs Wind Electric Company traveled to the antarctic in 1933 with famed explorer Rear Admiral Richard Byrd, where it ran problem-free for twenty-two years.

CHALLENGES FROM FOSSIL FUELS AND NUCLEAR ENERGY

Electric power lines began to reach rural farms and ranches in 1936, when President Franklin D. Roosevelt signed the Rural Electrification Act (REA) into law. Electricity generated by an abundant supply of fossil fuels took over much of the wind charger's work. The small wind generators couldn't compete with the Rural Electrification Administration, which provided government subsidies, or gifts of money, for central power systems. The work of the Rural Electrification Administration also discouraged independent wind energy systems, which didn't fit into the main distribution grid.

Palmer Putnam decided the solution was to build a huge, high-speed wind machine that could serve an entire town by feeding electricity into an existing utility grid. In 1941, Putnam, an engineer, constructed the Smith-Putnam wind turbine. It was a 250-ton wind generator standing more than 100 feet high. Built on a windy mountain in Vermont, the Smith-Putnam turbine operated for approximately sixteen months before it experienced mechanical failure. Without financial support from either the government or private industry to make the necessary repairs, the turbine was discarded. Putnam's efforts were not wasted, however. His innovative turbine laid the foundation for all subsequent wind energy projects.

The 1950s and the 1960s marked the birth of the nuclear age and America's belief that atomic energy would eventually replace the need for fossil fuels. While the United States took a break from wind energy development, across the Atlantic Ocean, wind turbines were making their debut.

Denmark Powers Ahead

Denmark took the lead in developing wind-power technologies with the Gedser wind turbine. Efficient and reliable, it ran without fail from 1957 to 1967.

Several factors renewed America's interest in developing wind energy as an alternative source of electrical power. One was that nuclear energy was proving to be costly and unsafe. Another

A windmill stands near a power plant in Outlaw Trail, Arizona. By the mid twentieth century, fossil fuel had replaced wind power as the primary source of energy.

was the rise of the environmental movement, which supported alternative energy sources. Perhaps the most significant factor was the oil shortage in the United States, which began in 1973. The United States, as well as countries around the globe, added wind turbines to their energy programs in an effort to reduce their dependence on oil. Foreign oil was expensive and new domestic oil drilling would be environmentally destructive.

Between 1973 and 1988, the U.S. federal wind-power program spent $380 million mostly in pursuit of large wind turbine development. The assumption was that the bigger the machine, the greater the energy output and the lower the costs. Unfortunately, mechanical failures made the large wind turbines unreliable.

Denmark, on the other hand, had designed a simpler yet far superior machine. The Danes' medium-sized but heavyweight turbines were, above all, reliable. And they dominated the wind market in California, where most of the turbines were erected during this era.

WIND POWER MAKES A COMEBACK IN THE UNITED STATES

Despite the failed efforts in the United States, federal and state tax credits helped the wind energy business to grow substantially in the first half of the 1980s. The opportunities to harvest wind multiplied, especially with the passage in 1978 of the Public Utility Regulatory Policy Act (PURPA). This federal legislation required utility companies to purchase electricity from small producers, such as wind companies, enabling the wind industry to reach more customers. Private investors rushed in to be part of a growing market.

After 1985, however, the tax incentives for wind power started to dry up and funding was cut. Quite a few wind companies went out of business. To make matters worse, many of Denmark's popular turbines, which were experiencing structural problems, had to be shut down.

A group of three-bladed wind turbines line a wind farm in Fyn, Denmark. Denmark is a world leader in the development of wind-power technologies.

Throughout the 1990s, the wind industry fought to be seen as a worthy, dependable source of energy. Fortunately, new government policies were created to encourage clean, renewable energy sources. Wind companies rushed to design, test, and manufacture turbines that would be durable and cost-effective. Between 1998 and 1999, wind power production in the United States rose 29 percent, according to World Watch Institute, which monitors environmental trends.

During most of the nineties, in fact, wind energy was the world's fastest-growing source of energy. It remains so today. The nearly fourfold growth in world wind power use over the last five years is matched only by the growth in the computer industry.

3 THE BUSINESS OF WIND ENERGY

One of the marvels of wind is that it is free. It is an invisible force that can be captured but not hoarded by anyone who wants to use it.

Wind plants may be owned by a local electric utility or by any other company in the business of generating electricity. Today, the U.S. electric industry is being restructured in many states to encourage more competition among organizations that generate energy. As part of the restructuring, consumers now have a choice of electricity providers.

BUYING WIND POWER

Increasingly, electric companies, or utilities, buying wind power from private corporations that own wind plants, are selling wind energy to consumers as part of their "green power" programs. Green power refers to energy produced from clean, renewable resources, including wind.

Utilities in many states also have net metering programs to make it affordable for homes and businesses to produce some of their own electric power using wind and other renewable sources of energy. When customers need more electricity than their wind turbines can produce, they buy it from the utility grid. When the wind turbines produce more electricity than is needed, the excess is sent to the utility company and, in effect, banked for the customers' future use. This lowers the customers' electric bills and helps them avoid the need to install expensive battery storage systems.

Wind plants are often built on leased land. Farmers and ranchers are increasingly allowing developers to build turbines on their vast acreage, earning a profitable second income without disturbing their livestock or crops. A typical leasing arrangement provides a landowner with an income of approximately $2,000 per year from just one turbine connected to a utility grid, while removing only about an acre from agricultural production.

In some cases, wind farms are built on city-owned land. Leasing the land to wind energy developers can generate substantial income for a city, bringing money to the local community by providing tax revenues and reducing the need to import costly coal and gas. Wind farms also provide jobs, particularly on the Great Plains, which has been called the Saudi Arabia of wind power. At the present time, the U.S. wind industry employs more than 2,000 people directly and contributes to the economies of forty-six states.

BUILDING WIND PLANTS

Farmers like William and Roger Kass of Woodstock, Minnesota, add to their farming income by renting out space on their land for wind turbines.

Before the wind plant can be constructed, a developer must find a suitable location. The job of finding an area that meets the requirements of a wind power plant is called siting. Developers must study wind speeds at a prospective site for a full year, negotiate a land lease with property owners, analyze access to local roads that can handle the transport of heavy industrial equipment, and consider issues such as environmental concerns and public grievances. It takes a total of two years on average to complete the planning and the business development, and to obtain permits. It takes another four to six months to construct the

Windsmiths attach a 100-ton rotor to a wind turbine in Goodnoe Hills, Washington.

wind plant. Wind turbine manufacturers make and sell equipment to developers and utility companies.

CONSTRUCTION SPECIALISTS

Engineers, machinists, and welders work on turbine design and construct the turbines. Computer specialists are required to handle the turbines' complex software system. Field maintenance workers, called windsmiths, inspect the turbines regularly and make repairs. It is a difficult, even a dangerous, job that requires climbing tall towers in all types of weather.

INVESTING IN WIND FARMS

Building a wind farm is a big investment. To install twenty-six turbines with the capacity to generate 750 kilowatts of electricity each, developers can expect to spend approximately $20 million. Government incentives can reduce costs and boost industry growth. In the United States, financial and technical support is provided through a program run by the Department of Energy (DOE),

DID YOU KNOW ?

In 2001, close to $375 million in research and development funding went to renewable energy technologies. Of that amount, $40 million was invested in wind-energy systems.

called the Wind Energy Program. Support also comes from the National Renewable Energy Laboratory's National Wind Technology Center, and Sandia National Laboratories.

HIGH-PERFORMANCE TURBINE TECHNOLOGY

National laboratories are involved in a turbine research program to assist the U.S. industry in developing high-performance wind turbines that can compete for global energy markets. The idea is to design commercially viable large-scale wind-turbine systems as well as small ones.

The federal wind-energy program has three goals. The first is to develop advanced wind-turbine technologies capable of reducing the cost of energy from wind to two cents per kilowatt-hour in fifteen-mile-per-hour winds. The second goal is to establish the U.S. wind industry as an international technology leader by capturing 25 percent of world markets. The third goal is to increase wind-powered generating capacity in the United States enough to meet the electrical needs of 2.5 to 3 million households.

A federal policy now provides a production tax credit for electricity generated by a wind plant during its first ten years of operation. This helps the wind industry compete with other energy industries that receive billions of dollars in federal subsidies each year.

Some state governments offer incentives to wind companies as well. For instance, Minnesota provides developers with tax exemptions. State governments are beginning to mandate the development of renewable energy. In 1999, a law was signed in Texas that requires each electricity supplier or generator to provide at least 10 percent of its electricity from renewable energy sources. As a result, the construction of 2,000 megawatts of new renewable energy generators (including wind power) are planned for the state by 2009.

At the beginning of 2000, plants within the United States could generate approximately 2,000 megawatts of wind power. By the end of 2001, at least forty U.S. wind-energy projects in twenty states were operational, boosting capacity to more than 4,000 megawatts, enough to supply about 1.2 million homes. The United States is now the second largest wind market in the world.

WIND-POWER LEADERS

Germany is the leading wind-power market, producing more than 6,000 megawatts of capacity as of the end of 2000. Third-ranked Denmark currently generates 15 percent of its electricity from wind, and aims to produce 50 percent of its electricity from wind, as well as other renewables, by 2030.

Among developing nations, India and China have the largest amount of installed wind-energy-generating capacity. India, in fact, is ranked fifth behind Spain, and India's government is considering policies to promote renewable sources of energy including wind power. China received a $12 million grant from the Global Environment Fund to aid in the development of wind-power generation and to reduce emissions of greenhouse gases. Morocco opened its first wind-power plant in 2000. The 50-megawatt wind plant is expected to generate an amount equal to 2 percent of Morocco's total electricity generation.

Global market opportunities for those in the business of wind power are tremendous. Sales of wind power plants, village power systems, wind-electric water pumping systems, wind-power support for telecommunications, and small turbines for homes, businesses, health clinics, and community centers could translate into billions of dollars in sales.

In fact, the U.S. wind industry is already marketing utility-scale wind turbines and complete power plants around the world. Since 1990, U.S. firms have installed wind turbines in Canada, the Netherlands, Mexico, South America, Spain, Ukraine, and the United Kingdom. Clearly, the wind business is soaring in the United States and globally as countries discover the power of the wind.

Wind power is safe, clean, and nonpolluting. It does not damage our air or our water. In fact, we can lower air pollution and the release of carbon dioxide, the leading greenhouse gas, by simply replacing fossil fuel power plants with power plants that use wind to generate electricity.

SAVING THE ENVIRONMENT

A single 750-kilowatt (kW) wind turbine engine prevents as much carbon dioxide emission each year as could be absorbed by 500 acres of forest. Less pollution means fewer health and environmental problems.

Water conservation is another plus for wind. While other power plants use large amounts of water for operation and maintenance, wind turbines consume only a little water to clean rotor blades in dry climates where rainfall does not wash away dust and insect buildup.

Performance of Wind Turbines

Because wind velocity changes frequently, wind turbines cannot operate at full peak—100 percent capacity—very often. The capacity factor is a measure of the number of hours in the year that turbines operate at full peak. The capacity factor of wind turbines is about 35 percent. The capacity factor of coal-fired power plants is about 80 percent. As engineers design better ways to capture wind, the turbines will perform better.

Maintenance

Wind technology has a long life span. Many American farm windmills have been used continuously for generations. Some traditional European windmills have been working for almost 300 years.

Many wind turbines have generated electricity since the early 1980s, and turbine blades commonly last from ten to twenty years, or longer. What's more, wind farms almost never need to be entirely shut down for maintenance and repairs because when one turbine is being repaired, the others can still generate electricity. For coal or nuclear power, shutdowns often involve the entire plant, sometimes for long periods. This is why having a diverse mix of energy sources is vital.

WIND POWER AND THE U.S. ECONOMY

Wind energy is a resource we can access right here in the United States. We don't have to pay hefty fees to import it as we do for for-

A maintenance worker prepares to work on a wind turbine tower. One advantage that wind farms have over power plants that use fossil fuels is that they are rarely shut down completely for maintenance and repairs.

eign oil. The United States, dependent on Middle Eastern countries for oil, is always at risk of having the supply cut off. Using wind power will help to make us self-sufficient.

According to the U.S. Department of Energy, wind technology provides more jobs per dollar invested than does any other energy technology—more than five times that from coal or nuclear power!

It takes two and a half years to plan and build a wind plant. That's faster than it takes to construct a coal-fired plant or a nuclear plant, but longer than it takes to complete a gas-fired project. Although fossil fuel generators cost less to build, they are more expensive over the long haul because of higher fuel and operating expenses, as well as the hefty environmental and health costs. Furthermore, the average energy payback period for a wind plant is a matter of months. (The energy payback period refers to how long the plant must work to generate as much energy as was spent to build it.)

THE COST OF WIND POWER

It's essential to the future of wind power for it to cost less than other energy sources. Currently, the average price range for wind energy is three to six cents per kilowatt-hour, which is comparable to burning fossil fuels. As technology advances, the price is expected to drop.

SMALL WIND FARMS

The business side of wind power can be troubling for small-scale wind farms. Even though wind farmers quickly show a profit, many banks are hesitant to provide them loans because they are unfamiliar with the industry and consider it risky. When financing is provided, bankers charge higher interest rates than they do for other energy projects. Wind farmers still struggle with how to sell wind energy profitably to local utility companies.

THE POTENTIAL OF WIND POWER

If wind power were given the chance to reach its full potential, its contribution to energy production would be huge. According to a recent study by the Pacific Northwest Laboratory for the U.S. Department of Energy, 0.6 percent of the continental United States's land area (16,000 square miles) could supply 20 percent of the country's electricity using current technology. Wind turbines, equipment, and access roads would use less than 5 percent of that land. The remaining land would be primarily for farming and ranching.

The same study found that if 13 percent of the land in the continental United States were used to generate wind power, it would produce triple the amount of energy we currently consume using fossil fuels. But the wind power industry is limited for now. The biggest obstacle to its growth seems to be transmission—getting the power to the people.

CHALLENGES

Wind turbines must be placed where there's plenty of wind. Many of the world's windiest areas are located in remote places, making it costly to transmit the power to heavily populated areas. In the United States, most transmission lines outside cities were installed decades ago and can't transmit much more power. The lines can become overloaded when additional power is fed into them, shutting down the whole system. New and improved power lines would solve the problem, but they are expensive.

Another hurdle for wind developers is finding sites that don't disturb protected areas like wildlife preserves. Proper turbine placement avoids bird sanctuaries and the routes of migrating birds which might fly into spinning blades. To ensure bird safety, newer towers are closed tubes rather than the open, lattice style of past models that attracted bird nesting. The slower-moving, more visible blades of today's large turbines should also reduce the number of bird collisions.

Public Opinion

Public acceptance is another issue. Not everyone likes the look of wind farms, saying that they disturb the natural beauty of the landscape. There have also been reports of interference with radio and TV signals as well as complaints about noise. Again, careful siting— away from people—can overcome these concerns. Furthermore, it is downwind machines whose motors can produce pulsating sounds that annoy some people. The vast majority of large wind turbines have an upwind design that enable the rotors to face the wind. The whirring sound of upwind machines is muffled by the wind.

Noise is more of an issue with small wind turbines because when the wind gets too strong, the blades start to flutter, slowing down the turbine. The fluttering can sound like a chainsaw, but only for short bursts. Southwest Windpower, a manufacturer of small turbines, is introducing a redesigned machine that brakes itself electronically instead of relying on the blades. A quieter operation should result.

This photo shows a wind turbine tower under construction in Texico, New Mexico. Wind power companies can minimize the criticisms of opponents by erecting their wind farms in places where they are out of sight and hearing, and where they are less likely to interfere with broadcast signals.

The wind may be invisible, but its advantages are clear to see. As the public learns of its virtues, and participates in the decision-making process for wind-power development, wind farms will likely gain widespread support and become a common sight on our landscape.

One of the world's largest wind power plants will soon become fully operational. A 300-megawatt wind farm along the Oregon-Washington border, it will serve about 70,000 homes. In addition, the Bonneville Power Administration, a Department of Energy federal agency, signed agreements with private firms for the development of seven wind power projects. The projects will provide 830 megawatts of electricity to the western region of the country, where power shortages have occurred in recent years. More than 700 megawatts of wind-generated energy are now being developed in Texas, too.

ON THE DRAWING BOARD

In the planning stages is a massive 3,000-megawatt wind farm in South Dakota named Rolling Thunder which will feed power to the mid-western region around Chicago. This proposed operation is one of the largest energy projects of any kind in the world today.

Meanwhile, researchers wrestle with the challenge of perfecting the technology so that wind energy can be harvested and delivered reliably to consumers at competitive prices. If the price of wind power drops 30–50 percent, as researchers believe it will, utility-scale wind systems will be one of the most inexpensive forms of electric generation in this century.

Large cost reductions are possible, researchers explain, because right now wind turbines are heavier and costlier than they need to be. They are overdesigned because we cannot yet accurately predict how much force they will have to withstand. Forces, or loads, on a wind turbine come from gusts of wind or turbulence that can damage the parts. As we learn to predict the force of the wind more accurately, we can build lighter, less expensive machines.

WIND-TUNNEL EXPERIMENTS

Experiments in a wind tunnel—basically, a large room with huge fans to mimic the wind in a controlled environment—help engineers more accurately predict the stresses placed on wind turbines during

Scientists look at six forty-foot fans at a NASA wind tunnel in Moffett Field, California.

operation. The collected data will help in the design of rotor hubs that are more flexible and lighter weight. Not only will the new technology cost less, but, as an added bonus, it will capture more wind energy!

New Blade Technologies

New blade technologies will further improve reliability and reduce expenses. For example, old blade materials may be replaced by carbon or glass-carbon hybrids. Adaptive blades are another amazing development. By changing shape in response to the wind, adaptive blades could increase turbine performance by as much as 35 percent.

COMPUTERIZED MAPPING AND FORECASTING

Critical to wind-energy development is the ability to estimate how much wind energy is available at a potential site. Work continues on developing a computerized mapping system that will incorporate new meteorological, geographical, and terrain data to make those estimates more accurate.

And what if we knew precisely when the wind would blow? Wind-forecasting systems are being tested for their ability to make this prediction accurately. By knowing how much wind-generated electricity can be produced at a particular time, utilities would be better able to schedule power generation to meet consumers' needs for electricity. It would also allow electricity sellers to commit a certain amount of power at specific times and to receive a fair price for their product.

WINDS OF THE FUTURE

In an effort to increase the efficiency of wind turbines and to lower the price of the electricity they produce, larger turbines will be built. As we've learned, longer rotor blades generate more electricity. Likewise, taller towers can reach up to where there's more wind. In the next few years, average tower heights are predicted to rise from 100 feet to approximately 230 feet.

Brothers Larry and Roy Josoff stand in front of a vertical-axis wind machine they have invented. The machine, which uses sails instead of propellers, can generate electricity using less wind power than is needed to generate electricity by propeller-driven turbines.

In 2000, manufacturers began to install machines that are 250 feet high and generate between one and two megawatts. In the future, offshore turbines as tall as 500 feet will generate even more power! These giants will be installed primarily off Europe's coastlines.

FLOATING AND FLYING WIND FARMS

The floating wind farm, although it exists only as a concept, has enormous potential for large-scale energy production. The floating systems would make use of the vast amount of offshore area that is too deep for bottom-mounted systems (greater than 60 feet) but shallow enough (less than 1,200 feet) for a structure that floats. To be kept in place with lines and anchors, the technology is comparable to floating, offshore oil- and gas-drilling platforms. By locating the turbines farther from shore, where the winds are generally higher, floating wind farms will be able to put out tremendous amounts of energy.

If a wind farm can float, can it also fly? One Australian researcher recently concocted an idea for a flying windmill called a gyromill, which looks like a giant kite with rotor blades. As the blades of the gyromill turn in the wind, it will generate electricity that reaches the ground through a copper cable. Whether it's practical remains to be seen.

SMALL WIND SYSTEMS

The proven technology of small wind systems, on the other hand, will continue to spur wind power development both here and abroad. In the United States, 24 percent of the population live in rural areas, with a growing number of people living in remote areas not served by electric power lines. Around the world, an estimated two billion people do not yet have access to electricity. In many of these places, traditional methods of electricity generation are costly and produce pollution. Several U.S. companies are now beginning to supply the global marketplace with small off-grid wind systems that can operate in remote locations with low to moderate wind speeds. For instance, Southwest Windpower has created a new wind-electric water pumping system that will help farmers in Mexico to pump larger amounts of water than is now possible using traditional mechanical water pumping systems.

A United Nations Development Programme project underway in a poor agricultural region of Brazil will help fund the construction of wind and solar equipment to supply the area with water and electricity. In this way, renewable energy can improve the health and productivity of isolated people.

Hybrid Systems

Research at the National Wind Technology Center includes developing wind hybrid systems that combine wind energy with other energy sources, such as solar cells and diesel or gas generators, to serve Third World villages and other small, isolated communities. The research will give the wind industry the opportunity to bring new hybrid technologies to market. "Mini-grid" village hybrid systems are often more economical than installing power lines for communities in remote but windy regions.

Other hybrid systems are also being proposed, such as one that combines the power of wind and the burning of biomass, specifically garbage. When the wind is too slow to rotate the turbines, biomass-burning furnaces would pick up the slack. So far, the technology is

In the future, American homes may be powered by fuel cells that operate on hydrogen, which can be derived from wind energy.

too expensive, with too many problems, to pursue. But it does offer a compelling way to deal with waste disposal while lowering our dependence on fossil fuels.

FUEL CELLS

Perhaps the fuel cell holds the greatest promise for the growth of wind power. A fuel cell is a device that generates electricity by combining air or oxygen with the energy of a fuel such as hydrogen, natural gas, or gasoline. When fueled with pure hydrogen, fuel cells create energy that is free of pollution. Today, hydrogen is mostly produced from natural gas. In the future, the preferred way to produce hydrogen may be through electrolysis, using an electrical current to split water into hydrogen and oxygen. The electrolysis would occur using renewable energy sources such as wind power.

The energy from fuel cells can power many things, from cell phones and city transit buses to electrical plants and U.S. space shuttles. The biggest opportunity to eliminate pollution will come when cars run with energy efficient fuel cells powered by hydrogen. The cars of tomorrow may be fueled by hydrogen produced at wind farms and piped across the country to your local gas station.

Whichever direction it takes, the wind will remain one of our most abundant and economical renewable resources. And, as the technology advances, wind power will guide our future in ways we are only now beginning to discover.

GLOSSARY

acid rain Rain that becomes toxic from air pollutants produced by burning fossil fuels.

airfoil Curve-shaped body designed to create aerodynamic lift and improve turbine performance.

anemometer A device to measure wind speed.

capacity factor A measurement of how much time a wind turbine spends generating electricity.

diameter The length of a straight line through the center of an object.

downwind In the same direction as the wind is blowing.

fantail A propeller-type wind rotor mounted on the side or back of a windmill, or horizontal-axis turbine, that keeps the machine aimed into the wind.

fossil fuel Any fuel made from the decayed remains of ancient plant and animal life; includes coal, natural gas, and oil.

greenhouse effect The warming of Earth's climate, a result of gases in the atmosphere that trap the Sun's heat. Burning fossil fuels contributes to the gases.

grid utility distribution system The network that connects electricity generators to electricity users.

horizontal Extending in the same direction as the horizon.

horizontal axis wind turbine A machine with blades attached to a horizontal axis or support; the blades rotate vertically.

hybrid Someone or something whose background is a blend of cultures or technologies.

kilowatt (kW) A measure of power equal to 1,000 watts.

kilowatt-hour (kWh) A unit of energy that uses one kilowatt in one hour. Home electricity use is measured in kilowatt-hours.

megawatt (MW) A measure of power equal to one million watts.

nacelle The body of a propeller-type wind turbine, containing the gearbox, generator, blade hub, and other parts.

nonrenewable energy Energy from a source that is not easily replaced, such as fossil fuels and nuclear energy.

renewable energy Energy from a source that is easily replaced, such as wind power, solar power, water power, and biomass.

rotor The rotating part of a wind turbine, including either the blades and blade assembly or the rotating portion of a generator.

rotor diameter The diameter of the circle swept by the rotor.

turbulence The changes in wind speed and direction, frequently caused by obstacles.

upwind In the direction from which the wind is blowing.

vertical-axis turbine A machine with blades attached to a vertical axis or support; the blades rotate horizontally.

watt A unit of power used to measure electrical current.

wind farm A group of wind turbines, also called a wind power plant.

wind turbine A modern windmill that generates electricity.

FOR MORE INFORMATION

American Wind Energy Association
122 C Street NW, Suite 380
Washington, DC 20001
(202) 383-2500
Web site: http://www.awea.org

U.S. Department of Energy
Wind Energy Program
1000 Independence Avenue SW
Washington, DC 20585
Web site: http://www.eren.doe.gov/wind/

WEB SITES

Due to the changing nature of Internet links, the Rosen Publishing Group, Inc., has developed an online list of Web sites related to the subject of this book. This site is updated regularly. Please use this link to access the list:

http://www.rosenlinks.com/lfe/wind/

FOR FURTHER READING

Bailey, Donna. *Facts About Energy from Wind and Water*. Austin, TX: Raintree/Steck-Vaughn Co., 1991.

Challoner, Jack. *Energy*. New York: Dorling Kindersley, 1993.

Parker, Steve. *Electricity*. New York: Dorling Kindersley, 1992.

Rickard, Graham. *Wind Energy*. Milwaukee, WI: Gareth Stevens Children's Books, 1991.

Woelfe, Gretchen. *The Wind at Work: An Activity Guide to Windmills*. Chicago: Chicago Review Press, 1997.

BIBLIOGRAPHY

Cohn, Laura. "A Comeback for Nukes?" *Business Week*, April 23, 2001, pp. 38–41.

Eisenberg, Daniel. "Which State Is Next?" *Time*, January 29, 2001, pp. 45–48.

Epstein, Paul R. "Is Global Warming Harmful to Health?" *Scientific American*, August 2000, pp. 50–57.

Fetter, Steve. "Energy 2050." *The Bulletin of the Atomic Scientists*, July/August 2000, pp. 28–38.

Fuel & Power (Revised Encyclopedia of Science). New York: Macmillan Pub., 1997.

Gipe, Paul. *Wind Energy Basics: A Guide to Small and Micro Wind Systems*. White River Junction, VT: Chelsea Green Publishing Co., 1999.

Kluger, Jeffrey. "A Climate of Despair." *Time*, April 9, 2001.

Lemonick, Michael D. "Life in the Greenhouse." *Time*, April 9, 2001.

Manwell, James F. "Windmill." Microsoft Encarta Encyclopedia CD-ROM, 2000.

Park, Jack. *The Wind Power Book*. Palo Alto, CA: Cheshire Books, 1981.

Raeburn, Paul. "Don't Write Off Energy Conservation, Mr. Cheney." *Business Week*, May 14, 2001, p. 46.

Raloff, Janet. "Power Harvests: The Salvation of Many U.S. Farmers May Be Blowing in the Wind." *Science News*, July 21, 2001, Vol. 160, pp. 45–47.

Righter, Robert W. *Wind Energy in America: A History*. Norman, OK: University of Oklahoma Press, 1996.

Symonds, William C. "Trying to Break the Choke Hold." *Business Week*, January 22, 2001.

Wardell, Charles. "Blackout." *Popular Science*, May 2001, pp. 64–67.

Woelfle, Gretchen. *The Wind at Work*. Chicago: Chicago Review Press, 1997.

Worden, Amy. "Many Switch to Green Power." *Philadelphia Inquirer*, January 22, 2001.

INDEX

CREDITS